LET'S CODE!

CODING ALGORITHMS

BY SAMANTHA S. BELL

CONTENT CONSULTANT
DAVID A. LASH
ASSOCIATE PROFESSOR OF COMPUTER SCIENCE
AURORA UNIVERSITY

Kids Core
An Imprint of Abdo Publishing
abdobooks.com

abdobooks.com

Published by Abdo Publishing, a division of ABDO, PO Box 398166, Minneapolis, Minnesota 55439.

Printed in the United States of America, North Mankato, Minnesota.
102023
012024

THIS BOOK CONTAINS
RECYCLED MATERIALS

Cover Photo: Chay Tee/Shutterstock Images
Interior Photos: Ermolaev Alexander/Shutterstock Images, 4–5; Shutterstock Images, 7, 9, 16, 23, 26, 29 (top); Yuriy Golub/Shutterstock Images, 10, 28 (top); Ground Picture/Shutterstock Images, 12–13, 29 (bottom); Chay Tee/Shutterstock Images, 15, 28 (bottom); Diego Thomazini/Shutterstock Images, 18; Wavebreak Media/Shutterstock Images, 20–21; Red Line Editorial, 25

Editor: Katharine Hale
Series Designer: Katharine Hale

Library of Congress Control Number: 2023939664

Publisher's Cataloging-in-Publication Data

Names: Bell, Samantha S., author.
Title: Coding algorithms / by Samantha S. Bell
Description: Minneapolis, Minnesota: Abdo Publishing, 2024 | Series: Let's code! | Includes online resources and index.
Identifiers: ISBN 9781098292744 (lib. bdg.) | ISBN 9798384910688 (ebook)
Subjects: LCSH: Coding theory--Juvenile literature. | Computer programming--Juvenile literature. | Computers and children--Juvenile literature. | Algorithms--Juvenile literature.
Classification: DDC 005.1--dc23

CONTENTS

Libraries are a great place to do research.

CHAPTER 1

USING ALGORITHMS

Melissa walked into the school library. She needed to write a report about a rainforest animal. She saw Trey and Natalie sitting at the computers.

Melissa sat down beside them. "What are you writing about?" she asked.

"I'm going to write about anacondas," Trey said. "They're huge snakes!"

"I'm writing about sloths," Natalie said. "Which animal did you choose?"

"I haven't decided yet," Melissa answered. "What other animals live in the rainforest?"

Melissa opened a web browser. She clicked on the search bar. Then she typed in "rainforest animals." A few moments later, she had her results. Photos of animals lined the top of the screen. Then came a list of websites about rainforest animals.

"I like this pink dolphin," she said.

"A dolphin in the rainforest?" Natalie asked. "That doesn't seem right."

Amazon River dolphins are also called botos. Males can be pink.

"It's called the Amazon River dolphin," Melissa said. "I'm going to find out more about it." She clicked on the search bar again. She typed "Amazon River dolphin." More photos and websites appeared. This time, all of the photos showed the dolphin.

Following Directions

Melissa was able to search for dolphins on the internet because of algorithms. An algorithm is an exact list of instructions. Algorithms are used to solve problems or complete tasks. Search engines use them to help people find information online. These algorithms are used to

Algorithms Everywhere

People use algorithms every day. Any step-by-step instructions are an algorithm. This includes the steps for solving math problems or checking out a library book. Directions from one location to another make up an algorithm. Algorithms are also used in daily activities such as following a recipe to cook a meal.

Search engines use algorithms to display the correct results.

make the **software** that tells a computer how to find information.

Melissa typed in the search bar. Then the algorithm ran the words through a set of instructions. It told the search software which results were closest to the words. The algorithm then told the computer what results to display.

Algorithms are step-by-step instructions.

Algorithms direct computer software in many other ways. They tell software to guide emails to the correct inboxes. Social media sites show posts based on algorithms. Many websites would not work without them.

Explore Online

Visit the website below. What new information did you learn about algorithms that wasn't in Chapter One?

Algorithms

abdocorelibrary.com/coding-algorithms

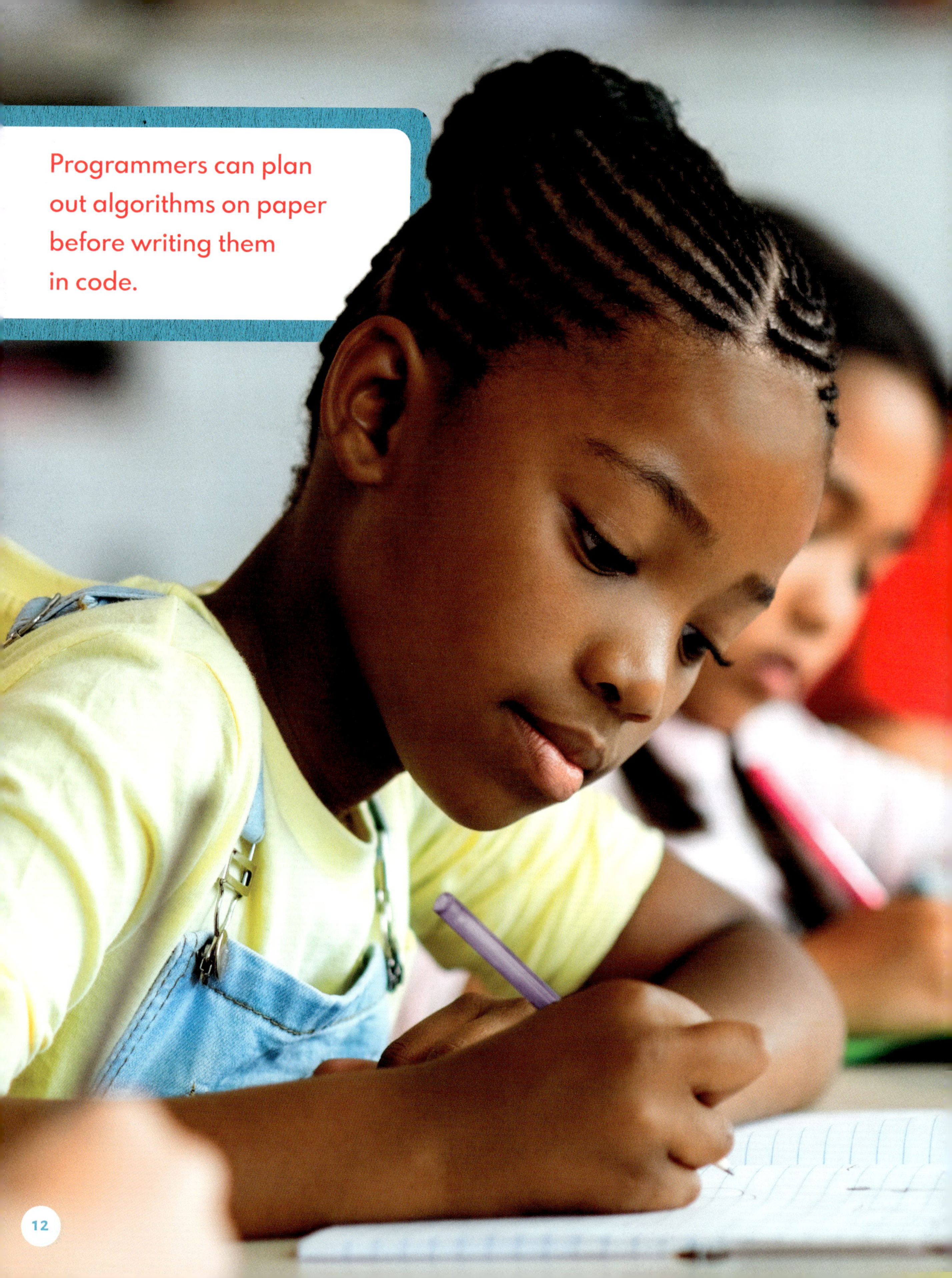

Programmers can plan out algorithms on paper before writing them in code.

CHAPTER 2

FROM INPUT TO OUTPUT

Writing algorithms means describing the steps the software will follow. Programmers can do this in different ways. They might write an algorithm as a chart or a graph. They could also write it in words.

Programmers write the planned algorithm in a **programming language**. Then the computer can understand it.

Every algorithm begins with input. This is the information the **user** puts into the program. A programmer codes the steps of the algorithm.

Ideas in the Making

Ada Lovelace was an English mathematician in the 1800s. She worked with an inventor named Charles Babbage. Babbage created the first design for an automatic computing machine. He believed it would be able to solve math problems. Lovelace published a paper about the machine. She showed how it could calculate a certain sequence of numbers. It was the first algorithm for a machine.

Algorithms need clear, step-by-step instructions in order to work correctly.

The programmer needs to make sure the steps are **precise** and complete. These instructions must include even the smallest actions. If a step is missing, the program may not work correctly.

Programmers must problem-solve if their algorithms do not work correctly.

The program follows the steps. Then it produces the output. The output is the solution to the problem. The algorithm may be very long. But every algorithm has an ending. The output is created by the last step.

Figuring It Out

Sometimes an algorithm does not get the expected output. This can happen for several reasons. There may be a problem with the algorithm itself. The input may also be wrong. Or one of the instructions may be incorrect or missing. Algorithms usually have multiple steps. The steps must be in the correct order. Otherwise, the algorithm will not produce the correct output.

Games such as Minecraft use algorithms to create worlds for players to explore.

For example, some video games use algorithms to create worlds. Players can explore these worlds. The worlds may have a mix of deserts, forests, and swamps. There may be rivers, lakes, and oceans. But the landscapes should be balanced. A world with all oceans and no land is not as fun for the player. Neither is a world with only deserts. A **faulty** algorithm could keep the worlds from being balanced.

There is often more than one correct algorithm for a problem. Some will solve the problem faster than others. The best algorithm is usually the one that works the fastest.

Further Evidence

Look at the website below. Does it give any new evidence to support Chapter Two?

All about Algorithms

abdocorelibrary.com/coding-algorithms

Search algorithms allow people to look up information on the internet.

CHAPTER 3

TYPES OF ALGORITHMS

There are many types of algorithms. Search algorithms use bots to help people find information online. Bots are programs that look through websites. They gather **data** about the sites they visit. The search algorithm uses this data.

When someone searches online, the algorithm knows which pages to show.

Sorting algorithms are another type. They arrange a list of items in a certain order. For example, they may be used to sort items in an online store. They could sort the items by price or popularity.

Most algorithms have steps where decisions must be made. There is more than one path to follow. The input and data decide which path the algorithm will take. For example, social media sites use these algorithms. They gather information about the user. Then they show posts the user will probably like. Algorithms for **streaming services** decide what content to recommend.

Streaming services use algorithms to suggest new movies or TV shows based on what a user has already watched.

Creating Algorithms

A programmer follows several steps to create an algorithm. First, a programmer describes the problem. They look at all the facts about it. Then they think about possible solutions. They decide which one would be the best.

Next, the programmer develops an algorithm. There are many ways to do this.

Using flowcharts is one of them. A flowchart uses shapes and symbols. These show the steps the algorithm will take. Some programmers use pseudocode. This is a description of what the algorithm must do. It is written in a natural language, such as English. Flowcharts and pseudocode help the programmer plan out each step. The programmer can then write the algorithm in a programming language.

Algorithm Engineers

Algorithm developers are also called algorithm engineers. Their job is to design and refine algorithms. They are often highly skilled programmers. They may also be mathematicians or scientists. Many algorithm engineers work on artificial intelligence (AI) systems. AI systems use the data they are given to learn and make decisions.

Planning a Program

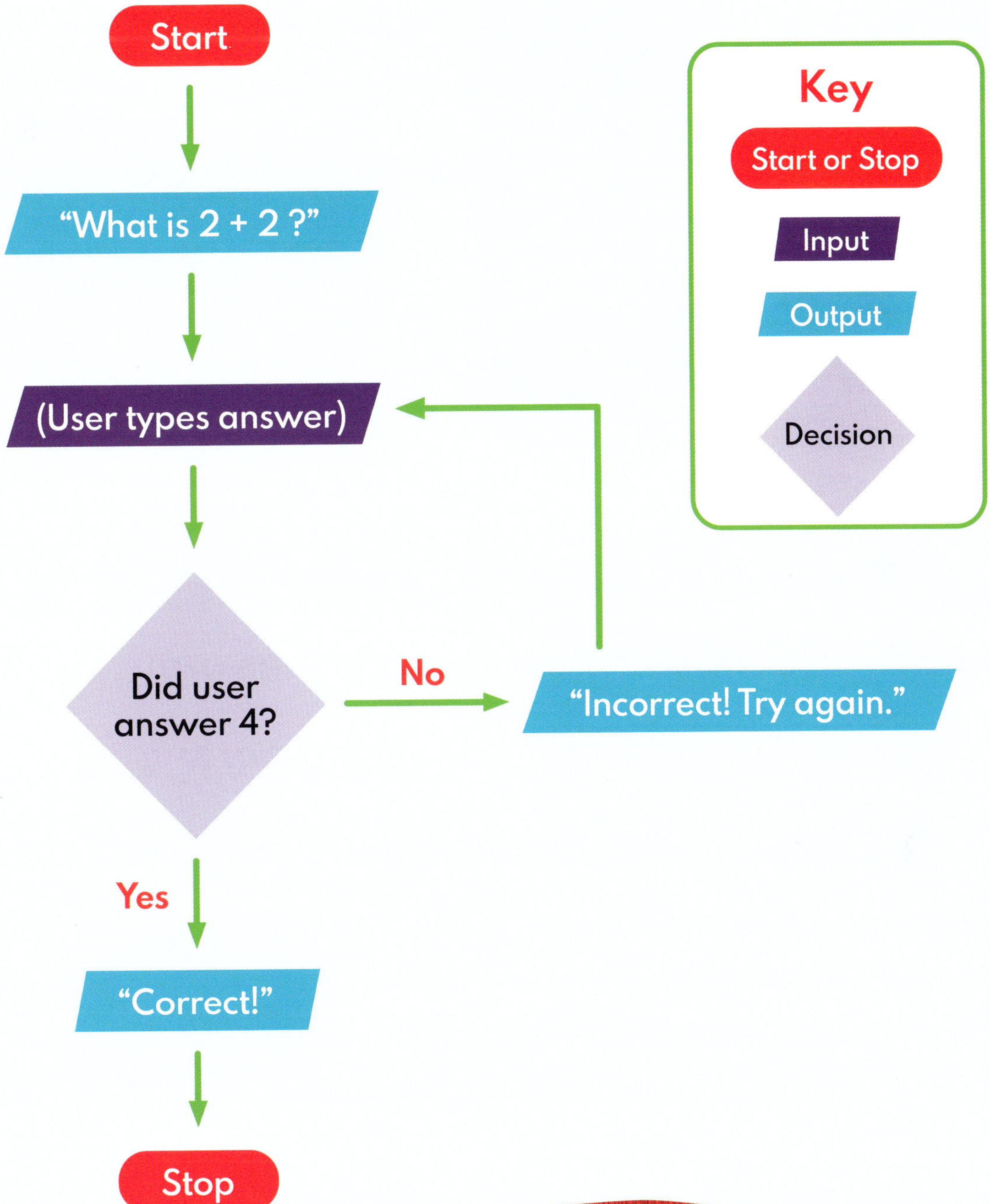

This flowchart shows the steps for an online quiz. A programmer could use the chart to write an algorithm.

Algorithms are an important part of coding.

Finally, the programmer reviews their work. They make sure the instructions are detailed. They check to see if the algorithm solves the problem. The programmer can try different inputs. The program must give the correct output every time.

Algorithms have become an important part of daily life. All computer programs depend on them. From the internet to video games, algorithms help things work better and faster.

PRIMARY SOURCE

Kjerstin Thorson is a professor. She described social media algorithms:

> The way it works is that you send different signals through your own behavior, so who you're friends with, what kinds of stuff you click on, what you read, what you look at, and that shapes what the algorithm thinks you want to see.

Source: Melissa Priebe. "The Social Media Sorting Hat." *MSU College of Communication Arts and Sciences*, 13 Aug. 2019, comartsci.msu.edu. Accessed 6 Apr. 2023.

Comparing Texts

Does this quote support the information in the chapter? Or does it give a different perspective?

CODING DATA

An algorithm is a set of step-by-step instructions that tells software how to solve a problem.

Algorithms begin with an input and end with an output.

There are many different types of computer algorithms, such as search algorithms and sorting algorithms.

Programmers can create algorithms in many ways, including charts, graphs, and written language. They convert these into programming languages.

Glossary

data
in programming, information that a computer or program can use

faulty
not working correctly

precise
communicated clearly and exactly

programming language
a vocabulary and set of rules for writing computer programs

software
all of the programs that give a computer instructions

streaming services
entertainment providers that deliver movies, TV shows, or music through an internet connection

user
a person who uses an electronic device

Online Resources

To learn more about algorithms, visit our free resource websites below.

Core Library CONNECTION
FREE! COMMON CORE MULTIMEDIA RESOURCES

Visit **abdocorelibrary.com** or scan this QR code for free Common Core resources for teachers and students, including vetted activities, multimedia, and booklinks, for deeper subject comprehension.

Booklinks NONFICTION NETWORK
FREE! ONLINE NONFICTION RESOURCES

Visit **abdobooklinks.com** or scan this QR code for free additional online weblinks for further learning. These links are routinely monitored and updated to provide the most current information available.

Learn More

Castaldo, Nancy. *Ada Lovelace*. DK Children, 2019.

González, Echo Elise. *Algorithms.* World Book, 2021.

Kulz, George Anthony. *Coding Functions.* Abdo, 2024.

Index

About the Author

Samantha Bell lives in the foothills of the Blue Ridge Mountains with her family and lots of cats. She has written more than 150 nonfiction books for students. She learned HTML through building websites and wants to learn more programming languages.